THE THREAT OF BILATERAL CLIMATE CHANGE

First Edition

Dr. Auke Schade

nemonik-thinking.org/

First Edition
Published 15 October 2016
@ nemonik-thinking.org
ISBN 978-0-473-37552-2

Abstract

This pilot study is a wakeup call. The international scientific-political establishment seems to accept that the recent increase in atmospheric carbon dioxide (CO_2) from about 280 to 400 part per million (ppm) has increased the global average atmospheric temperature to 1.4 °C above normal. Allegedly, this temperature is so high that the polar ice is melting, which in turn rises the sea levels flooding the low, fertile, industrial, and urban areas. This environmental disaster would force mass migrations threatening the global infrastructure and the survival of humanity. Even worse, this pilot study suggests that the thermal effect of CO_2 is ten times larger than accepted by the scientific-political establishment. Hence, their attempts to stabilize CO_2 at the current 400 ppm might not reduce the symptoms of climate change. Furthermore, it is proposed that we live in a glacial period, rather than in an interglacial period. Hence, CO_2 should be carefully managed in order to maintain interglacial temperatures. This would make CO_2 a beneficial substance, rather than a hazardous one. Due to inadequate datasets, further research of this topic is required urgently.

Dr. Auke Schade

My life started during the devastation of World War II. As a teenager, I worked as a carpenter and studied building engineering at night school. During the seventies, I became a financial manager for a multinational corporation, ran my own business, and studied economics in my spare time. My interest in the psychology of management extended to the interaction between the mind, body, and reality. In 1980, I immigrated to New Zealand where I obtained a doctorate in psychology from the University of Auckland. I consider myself to be a theoretical psychologist.[1] My mission is to make people the smartest thinkers they can be, which has led me to the development of nemonik thinking.[i]

Download free eBooks and videos
@ nemonik-thinking.org

---

i Appendix: Nemonik Thinking

nemonik-thinking.org/

Notes

Content

Notes

INTRODUCTION

The majority of the scientific-political establishment seems to accept that the recent human activity increased the atmospheric carbon dioxide (CO_2) concentration from about 280 to 400 parts per million (ppm). Allegedly, this sharp increase in CO_2 creates a global warming of about 1.4 °C above normal by trapping heat in the atmosphere.

Advocates of global warming argue that it is a direct and immediate threat to the survival of humanity. They point out that the resulting meltdown of the polar ice will rise significantly the sea levels. As a result, the low, fertile, industrial, and urban areas will be flooded. Among other consequences, the increasing heat will also change weather patterns, ocean flows, turn habitable areas into desserts, and decrease the global food supply.

The combined effects of global warming will force mass-migrations that will destroy a significant part of the global infrastructure. It is shown by the recent Syrian refugee crisis in Europe that humanity cannot cope adequately with such large-scale disasters. To limit the consequences of this manmade threat, international leaders have joined forces in an attempt to stop the increase of the atmospheric CO_2 concentration. However, if the current global temperature of 1.4 °C is already dangerously high, then stabilizing the CO_2 at the current concentration of 400 ppm will only delay the predicted global catastrophe.

The previous long delay in counteracting artificial global warming shows the cognitive inertia of the scientific-political establishment. After decades of hesitation and denial, they reluctantly agree to curb the atmospheric greenhouse gasses such as CO_2. However, if their decision is incorrect, then the cognitive inertia of that establishment, combined with the momentum of the consequences of their decisions, might not allow enough time for anymore corrections. Therefore, the essence of this pilot study is to accelerate the development of climatology.

Hegel's dialectic of knowledge implies that the development of science is a cognitive process, in which thesis, anti-thesis, and synthesis follow each other in an endless cycle. In accord, the nemonik accelerator increases the speed of scientific progress by fostering agreement during disagreement, while fostering disagreement during agreement (Schade, Think Smarter with Nemonik Thinking, 2016). Therefore, the widely accepted thesis of global warming is challenged by introducing the anti-thesis of bilateral climate change.

The unilateral hypothesis of climate change (global warming) holds that artificial greenhouse gasses, such as carbon dioxide (CO_2), increase the global temperature above normal by trapping heat in the atmosphere. In contrast, this study presents the bilateral hypothesis of climate change, which holds that the artificial global warming compensates for the natural global cooling that is caused by the hidden progression of a glacial period.[ii] The observed global average temperature of 1.4 °C is the balance of the opposing forces comprising artificial global warming and natural global cooling. This sensitive balance between those two polarities would explain the recent instability of the climate.

[ii] A hypothesis is an untested, but testable description of reality. Artificial refers to manmade phenomena and, therefore, artificial CO2 refers to domestic and industrial CO2.

Notes

ANALYSES

In this pilot study, temperature deviations are used, rather than absolute temperatures in order to foster comparisons across the planet. Hereafter, 'atmospheric temperature deviation' will be referred to as 'temperature' or 'TD'. The temperature will be expressed in degrees Celsius (°C) in reference to the mean temperature at Vostok from calendar year 989 to 1989 (Kyr1), which is adjusted to 0.0 °C (baseline). This baseline is valid and reliable, because the average temperature during the current interglacial from 7962 BC to 1989 equals 0.07 C.[2] Furthermore, 'atmospheric carbon dioxide concentration' will be referred to as 'CO_2', which will be expressed in parts per million (ppm).

Aim 1—This pilot study evaluates the unilateral and bilateral CO_2 hypotheses of climate change. For that reason, the recent short-term atmospheric data and the long-term atmospheric data will be subjected to statistical analyses.

The short-term dataset for the average global temperature (TDg) during the observation period from 1880 to 2014 was produced by (GISTEMP Team, 2015) and (Hansen, 2010).[3] The short-term dataset for the Hawaiian CO_2 (CO_2h) during the observation period from 1958 to 2014 was produced by (Mauna Loa Observatory (Scripps / NOAA / ESRL), 2015).[4]

Those short-term datasets were compared with the Vostok temperatures (TDv) from 422,766 years BP to calendar year 1989 and the Vostok CO2 (CO2v) from 414,085 to 2,342 years BP.[iii] BP means before the present calendar year 1989. Those long-term datasets were reconstructed from a 3,310 meters deep ice-core drilled in Vostok Antarctica (Petit, et al., 1999); (Petit, Vostok Ice Core Data for 420,000 Years, 2001); and (NOAA/NGDC Paleoclimatology Program).[5] The Antarctic atmosphere is among the driest on Earth and, therefore, humidity is less likely to confound the measurement of atmospheric temperatures. Hence, the Antarctic temperature is likely to provide a valid and reliable indication of the changes in the global temperature.

Unfortunately, the previously mentioned datasets are not perfect for the aim of this study. The data intervals, observation periods, and baselines within and between datasets differ considerably. The Vostok atmospheric data are reconstructed from the contents of the Antarctic ice and are long-term, while the global and Hawaiian data are short-term. Hence, the data have to be adjusted to foster comparisons, which might decrease the reliability of the resulting datasets by fostering regression towards the mean. Furthermore, the presented comparison of the global data with the local Antarctic data might decrease the va-

---

[iii] BP = Before Present. Present is the calendar year 1989.

lidity and reliability of the statistical analyses However, if the global data cannot be compared with local data, then the global temperature cannot be compared with the local Hawaiian CO2 data either. Hence, that objection would also cast doubt on the scientific basis for the widely accepted CO2 hypothesis and the decision to curb the greenhouse gasses. Therefore, further research is required urgently to create, extend, synchronize, and improve the datasets.

THE PREVIOUS 420 KYRS

TDv and CO2v across time

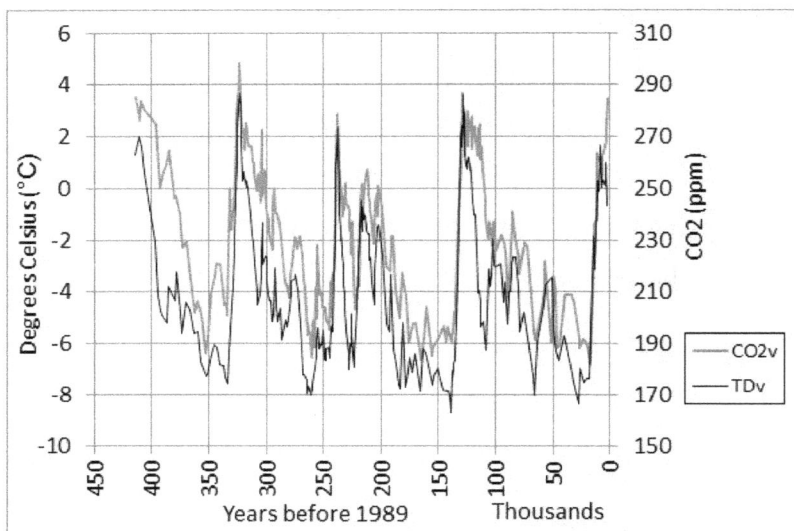

Figure 1—Vostok temperature (TDv) and Vostok CO2 (CO2v) from 414,085 to 2,342 BP with irregular data intervals.[6]

Correlation TDv and CO2v

Aim 2—Testing the hypothesis that the correlation between the Vostok temperatures (TDv) and the Vostok CO2 (CO2v) during the previous 420 Kyrs is statistically significant.[7][iv]

Result 1—A randomized probability test of the ranked correlations between TDv and CO2v, from 414,085 to 2,342 BP, yielded: $r = 0.87$; and $p = 0.0000$.[8] The result shows that $p < 0.05$ and, therefore, it supports Aim 2—Testing the hypothesis

---

[iv] Kyr = 1,000 years.

that the correlation between the Vostok temperatures (TDv) and the Vostok CO2 (CO2v) during the previous 420 Kyrs is statistically significant. [9][v] Nevertheless, during the period from 3,634 to 2,342 BP, the CO2v increased with 12 ppm, while TDv decreased with 1.7 °C.[10] This suggests that one or more missing variables are involved. A missing variable is a variable that is not included in the analysis.

Conclusion 1—The correlation between the Vostok temperature and CO2 during the previous 420 Kyrs is statistically significant. This suggests that both variables change in synchrony. Therefore, this result cannot reject the CO2 hypotheses of climate change. However, the result cannot prove those hypotheses either, because every correlation analysis is non-causal. A significant correlation cannot indicate whether temperature or CO2 is the cause of the synchronized changes. It is even possible that one or more missing variables are causing those changes. Therefore, the result provides non-conclusive support for both CO2 hypotheses of climate change.

---

[v] < means smaller than.

TDv as a function of CO2v

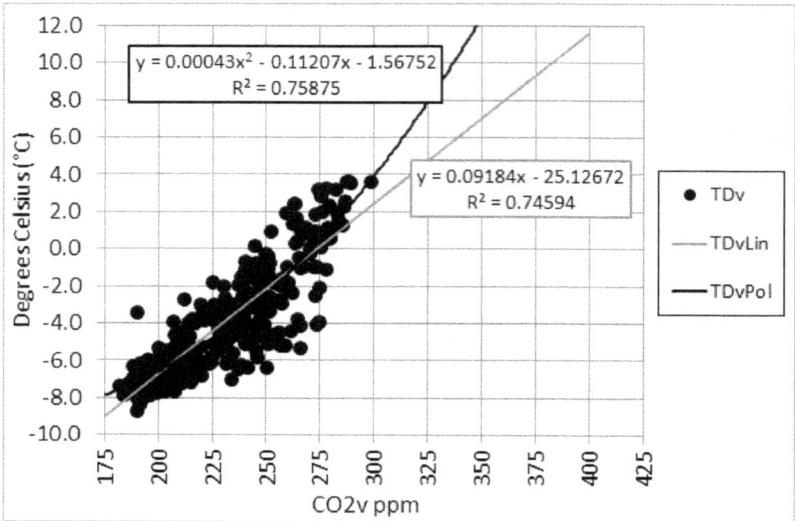

Figure 2—Vostok temperature (TDv) as a function of Vostok CO2 (CO2v) from 414,085 to 2,342 BP with irregular data intervals. TDvLin = linear function of TDv; and TDvPol = second order polynomial function of TDv.[11]

Aim 3—Deriving a mathematical function from the 420 Kyrs of Vostok atmospheric data in order to predict the Vostok temperature (TDv) and the Vostok CO2 (CO2v).

Equation 1 (Figure 2)—Describes the Vostok temperature TDv as a linear function of the Vostok CO2v (TDvLin): $y = 0.09184x - 25.12672$. In which: y = TDv and x = CO2v.[12] The linear function (TDvLin) and the second order polynomial function (TDvPol) shown in Figure 2 are both adequate, because they explain respectively 75% and 76% of the variance in TDv. Increasing the sensitivity of the polynomial function to the sixth order did not improve significantly the explained variance of TDvPol. The linear function was preferred, because it provides conservative predictions, optimizes the simplicity, and facilitates the comparison between the global and Vostok results.[13]

Result 2—Equation 1 predicts a temperature of 11.6 °C above the baseline at the current CO_2 of 399 ppm. On the other hand, Equation 1 predicts a temperature of -25.1 °C below the baseline at a CO_2 of 0 ppm.[14] Furthermore, Equation 1 predicts that the current CO_2 of 399 ppm has to be decreased with 125 ppm to 274 ppm in order to return the temperature to its baseline.[15]

Conclusion 2—Equation 1 is a reliable mathematical linear function predicting Vostok temperature (TDv) and Vostok CO_2 (CO2v).

Conclusion 3—At the current CO_2 of 399 ppm, the consequences of global warming are already substantial. Allegedly, polar ice deposits are melting and, therefore, sea levels are rising. Equation 1 suggests that the CO_2 has to be decreased to 274 ppm in order to return the temperature to its baseline. Hence, attempts of the scientific-political establishment to stabilize the CO_2 at the current level of 399 ppm are inadequate.

Interglacial thermal stability

An interglacial period is arbitrarily defined as a long-term climatological period with mean millennial temperatures equal or above the baseline. In contrast, a glacial period is a long-term climatological period with mean millennial temperatures below the baseline. The millennial temperatures were detrended in order to compare the interglacials of the previous 420 Kyrs without the confounding effect of the long-term cooling of the planet.

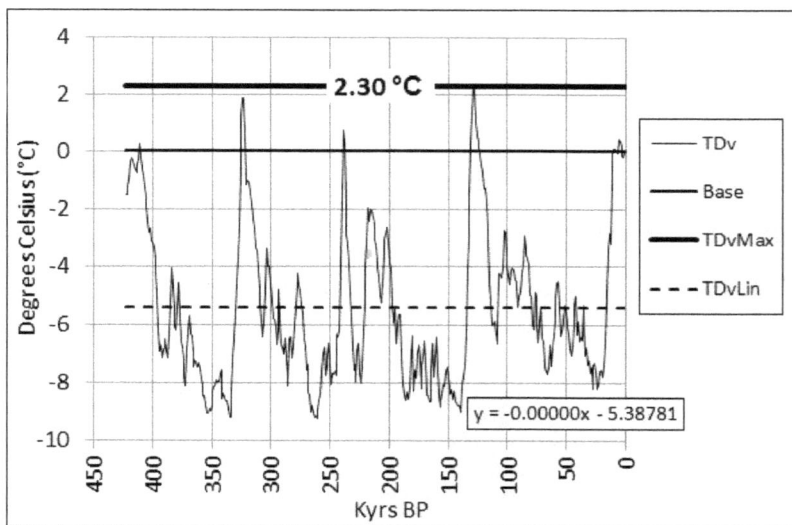

Figure 3—Vostok detrended millennial temperatures (TDv) from Kyr423 to Kyr1 BP.[16] Base = baseline; TDvMax = maximum of TDv; and TDvLin = linear function of TDv.

Aim 4—Testing the hypothesis that the Vostok thermal stability of the current interglacial period from Kyr10 to Kyr1 BP is significantly larger than the Vostok thermal stability during the observation period from Kyr423 to Kyr11 BP.[17]

Equation 2—The thermal stability index (TSI) is introduced, which equals: TSI = 1 / STD. In which, STD is the standard deviation of the Vostok detrended millennial temperatures (TDv) within a window of 10 Kyrs.[18] STD is a measure that changes in synchrony with the thermal instability. In contrast, TSI is a measure that changes in synchrony with the thermal stability.[19] To determine the TSI values within an observation period, the window slides across that period in steps of 1 Kyr.

Result 3—The mean TSI values obtained with sliding windows of 10 Kyrs yielded for the observation period from Kyr423 to Kyr11 BP the following statistics: minimum = 0.26; maximum = 4.02; mean = 1.33; STD = 0.72; and n = 404. In comparison, the TSI of the current interglacial period from Kyr10 to Kyr1 BP yielded the following statistics: TSI = 4.75; z-score = 4.72; and p = 0.0002.[20] The result shows that p < 0.05 and, therefore, it supports Aim 4—Testing the hypothesis that the Vostok thermal stability of the current interglacial period from Kyr10 to Kyr1 BP is significantly larger than the Vostok thermal stability during the observation period from Kyr423 to Kyr11 BP. In addition, the TSI of the current interglacial (4.75) is higher than the maximum TSI obtained with the sliding windows during the previous 404 Kyrs (4.02).[21]

Conclusion 4—The thermal stability of the current Vostok interglacial period from Kyr10 to Kyr1 BP is significantly larger than the thermal stability during the preceding observation period in Vostok from Kyr423 to Kyr11 BP.[22] This suggests that one or more missing variables confound the statistical comparison between the current interglacial and the preceding period. Therefore, the hypothesis is proposed that human activity has unknowingly stabilized the temperature of the current interglacial before the onset of the industrial revolution.

Interglacial duration

Aim 5—Testing the hypothesis that the duration of the current Vostok interglacial period of 9,926 years is significantly longer than the durations of the previous Vostok interglacial periods during the observation period from Kyr423 to Kyr11 BP.

Result 4—The start and finish of each interglacial was computed with linear interpolations of the adjacent millennial temperatures. The durations of the interglacials during the previous 420 Kyrs yielded the following statistics: minimum = 1,032; maximum = 7,682; mean = 3,575; STD = 3,053 years; and n = 4.[23] In comparison, the current interglacial yielded the following statistics: duration = 9,926 years; z-score = 2.08; $p = 0.0188$; and n = 4.[24] The result shows that $p < 0.05$ and, therefore, it supports Aim 5—Testing the hypothesis that the duration of the current Vostok interglacial period of 9,926 years is significantly longer than the durations of the previous Vostok interglacial periods during the observation period from Kyr423 to Kyr11 BP.

Conclusion 5—The current Vostok interglacial of 9,926 years is significantly longer than the Vostok interglacials during the previous 420 Kyrs. This suggests that one or more missing variables confound the statistical comparison. Therefore, in accord with Conclusion 4, it is proposed that human activity is that missing variable. In addition, the current temperature is above the baseline and, therefore, the current interglacial is still continuing. The significant long duration of the current interglacial supports the notion that the natural interglacial period has ended some time ago. In that case, we live already in a natural glacial climate that is warmed up by artificial CO_2 to an interglacial level. The precarious balance of the opposing natural and artificial forces could explain the recent instability of the climate.

THE PREVIOUS 135 YEARS

TDg and CO2h across time

Although, global temperatures were recorded since 1880, CO2 was only recorded in Hawaii since 1958.[25] Furthermore, calendar year 1958 was ignored, because it comprised incomplete data. Therefore, this evaluation is limited to the observation period from 1959 to 2014.

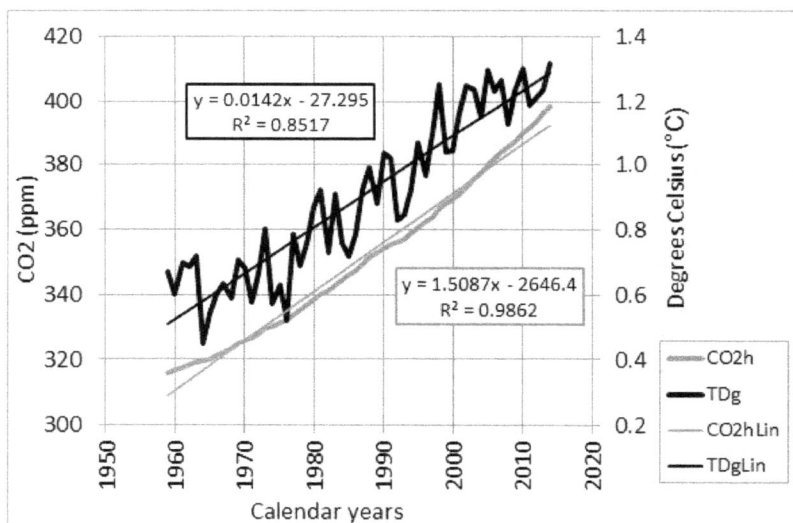

Figure 4—Global annual temperature (TDg) and Hawaiian annual CO2 (CO2h) from calendar year 1959 to 2014.[26] TDgLin = linear function of TDg; and CO2hLin = linear function of CO2h. TDgLin and CO2hLin are both adequate, because Figure 4 shows that they explain respectively 85% and 99% of the variances.

Correlation TDg and CO2h

Aim 6—Testing the hypothesis that the correlation between the average global temperatures (TDg) and the Hawaiian CO2 (CO2h) from 1959 to 2014 is statistically significant.[27]

Result 5—A randomized probability test of the correlation between the ranked TDg and CO2h values from 1959 to 2014 yielded: $r = 0.92$ and $p = 0.0000$.[28] The result shows that $p <$ 0.05 and, therefore, it supports Aim 6—Testing the hypothesis that the correlation between the average global temperatures (TDg) and the Hawaiian CO2 (CO2h) from 1959 to 2014 is statistically significant.

Conclusion 6—The correlation between the average global temperature and the Hawaiian CO2, from 1959 to 2014, is statistically significant. This supports the notion that both variables change in synchrony. Hence, this correlation analysis cannot reject the CO2 hypotheses of climate change. However, it cannot prove those hypotheses either, because every correlation analysis is non-causal. Therefore, the result provides only non-conclusive support for both CO2 hypotheses of climate change.

TDg as a function of CO2h

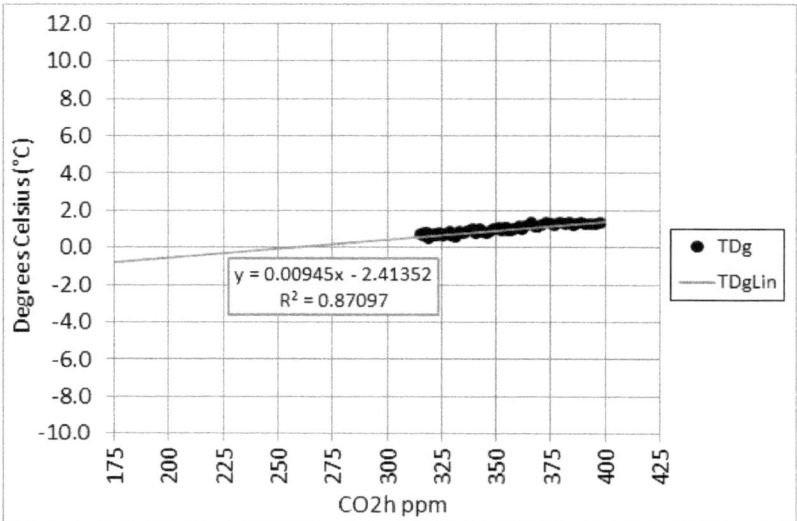

Figure 5—Global annual temperature (TDg) as a function of Hawaiian CO2 (CO2h) from calendar year 1959 to 2014.[29] TDgLin = linear function of TDg.

Aim 7—Deriving a mathematical function from the previous 55 years of global and Hawaiian data in order to predict the global temperature (TDg) and Hawaiian CO2 (CO2h).

Equation 3 (Figure 5)—Describes the global temperature TDg as a linear function of Hawaiian CO2h (TDgLin): $y = 0.00945x - 2.41352$. In which: y = TDg and x = CO2h. The linear function (TDgLin) is adequate, because Figure 5 shows that it explains 87% of the variance in TDg.

Result 6—Equation 3 predicts a temperature of 1.4 °C above the baseline at the current CO2 of 399 ppm. On the other hand, Equation 3 predicts a temperature of -2.4 °C below the baseline at a CO2 of 0 ppm.[30]   Furthermore, Equation 3 predicts that the current CO2 of 399 ppm has to be decreased with 144 ppm to 255 ppm in order to return the temperature to its baseline.[31]

Conclusion 7—Equation 3 is a reliable mathematical linear function predicting global temperature (TDg) and Hawaiian CO2 (CO2h).

Conclusion 8—Comparison of Equation 1 and Equation 3 shows that the Vostok thermal slope (TDvLin) is about ten times steeper than the global thermal slope (TDgLin). This suggests that the recent thermal effect of CO2 is much weaker than the thermal effect of CO2 during the previous 420 Kyrs. Climatologists have to explain this decrease in the thermal effect of CO2 in order to justify decisions about climate change.

Conclusion 9—At the current CO2 of 399 ppm, the consequences of global warming are already substantial. Allegedly, polar ice deposits are melting and, therefore, sea levels are rising. Equation 3 suggests that the CO2 has to be decreased to 255 ppm in order to return the temperature to its baseline. Hence, the recent attempts of the scientific-political establishment to stabilize the CO2 at the current level of 399 ppm are again inadequate.

CO2 HYPOTHESES OF CLIMATE CHANGE

Change in CO2

Aim 8—Evaluating whether the Hawaiian CO2 from 1959 to 2014 (COh) are significantly higher than the Vostok CO2 from 414,085 to 2,342 BP (CO2v).

Result 7—Visual inspection of the CO2 in [Figure 1—Vostok temperature (TDv) and Vostok CO2 (CO2v) from 414,085 to 2,342 BP with irregular data intervals.] and [Figure 4—Global annual temperature (TDg) and Hawaiian annual CO2 (CO2h) from calendar year 1959 to 2014.] seems to suggest that the current Hawaiian CO2 of 399 ppm is at an all–time high.[32] However, the data interval for the Vostok maximum CO2v of 299 ppm equals 681 years, while the data interval for the maximum CO2h of 399 ppm equals only one year.[33] Therefore, regression towards the mean is likely to confound the comparison.[34] For example, the mean of the CO2h from 1959 to 2014 equals 351 ppm.[35] Hence, increasing the data interval from 1 to 55 years decreases the mean CO2h from 399 to 351 ppm. Despite this large decrease, the Hawaiian data interval of 55 years is still much shorter than the Vostok data interval of 681 years.[36] At this stage, one can only speculate about the magnitude of the annual Vostok CO2 in the past. No one really knows. Therefore, regression towards the mean is still likely to confound the comparison between the short-term CO_2h and the long-term CO_2v.

Conclusion 10—Regression towards the mean is confounding the comparison between the annual Hawaiian CO_2 of 399 ppm in 2014 and the Vostok CO_2 from 414,085 to 2,342 BP. Hence, the current CO_2 of 399 ppm provides only non-conclusive support for both CO_2 hypotheses of climate change.

Change in Temperature

In order to account for different data intervals within and between datasets, the global temperatures (TDg) and the Vostok temperatures (TDv) were transformed into the means of window with durations of 50 years sliding across the observation periods in steps of 1 year (SW). Hence, this produced respectively the variables TDvSW and TDgSW.

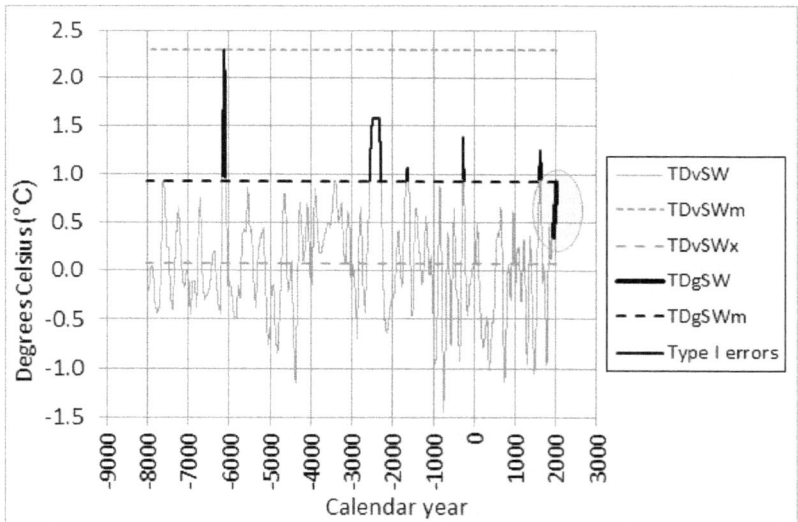

Figure 6—Vostok window temperatures from 7962 BC to 1928 and global window temperatures from 1929 to 2014.[37] SW = mean of a 50 years window that slides across the observation period in steps of 1 year; m = maximum; and x = mean.

Magnitude thermal increase

Aim 9—Testing the hypothesis that the global average temperature of 1.4 °C in 2014 (TDg) is significantly higher than the Vostok interglacial temperatures from 7962 BC to 1928 (TDv).[38]

Result 8—Figure 6 shows that the maximum global window temperature in 2014 (TDgSWm = 0.9 °C) is higher than the mean Vostok window temperature from 7962 BC to 1928 (TDvSWx = 0.1 °C).[39] However, during five separate occasions, the Vostok TDvSW values have been higher than the global thermal maximum (TDgSWm = 0.9 °C).[40] In addition, the Vostok thermal maximum (TDvSWm = 2.3 °C) is considerably higher than the global thermal maximum (TDgSWm = 0.9 °C).[41]

Result 9—If the temperature is below its baseline, then there is by definition no global warming. Hence, the evaluation of global warming makes only sense if the temperature is above the baseline.[42] In accord, during 4,671 of the 9,891 observation years the Vostok interglacial window temperatures (TDvSW) were higher than the baseline.[43] In addition, during 518 of those 4,671 decision years, the TDvSW values exceeded the global thermal maximum (TDgSWm = 0.9 °C).[44] Hence, p = 518 / 4671 = 0.11.[45] The result shows that p > 0.05 and, therefore, it rejects Aim 9—Testing the hypothesis that the global average temperature of **1.4** °C in 2014 (TDg) is significantly higher than the Vostok interglacial temperatures from 7962 BC to 1928 (TDv).[vi]

Conclusion 11—The results reject the hypothesis that the average global temperature in 2014 is significantly higher than the Vostok interglacial temperatures from 7962 BC to 1928.

[vi] > means larger than.

Slope thermal increase

Aim 10—Testing the hypothesis that the maximum slope of the uninterrupted increases in global temperatures from 1880 to 2014 (TDg) is significantly steeper than the slopes of the interrupted increases in the Vostok interglacial temperatures from 7962 BC to 1879 (TDv).[46]

Result 10—The slopes of 26 of the 51 increases in the Vostok interglacial window temperatures were steeper than the maximum slope of the global window temperatures of 0.00858 °C/year. Hence, the probability equals: $p = 26 / 51 = 0.51$.[47] The result shows that $p > 0.05$ and, therefore, it rejects Aim 10—Testing the hypothesis that the maximum slope of the uninterrupted increases in global temperatures from 1880 to 2014 (TDg) is significantly steeper than the slopes of the interrupted increases in the Vostok interglacial temperatures from 7962 BC to 1879 (TDv). In addition, the mean slope of the increases in Vostok temperatures of 0.01017 °C/year is steeper than the maximum slope of the increases in global temperatures of 0.00858 °C/year.

Conclusion 12—The results reject the hypothesis that the maximum slope of the uninterrupted increases in the global temperature from 1880 to 2014 is significantly steeper than the slopes of the uninterrupted increases in the Vostok temperatures from 7962 BC to 1879. The slope of the recent increase in global temperature is even below the natural average.

Duration thermal increase

Aim 11—Testing the hypothesis that the maximum duration of the uninterrupted increases in global temperature from 1880 to 2014 (TDg) is significantly longer than the durations of the uninterrupted increases in Vostok temperatures from 7962 BC to 1879 (TDv). [48]

Result 11—Of the 51 durations of the uninterrupted increases in the Vostok temperatures, 50 lasted longer than the maximum duration of the uninterrupted increases in the global temperature of 37 years. Hence, the probability equals: $p = 50 / 51 = 0.98$.[49] The result shows that $p > 0.05$ and, therefore, it rejects Aim 11—Testing the hypothesis that the maximum duration of the uninterrupted increases in global temperature from 1880 to 2014 (TDg) is significantly longer than the durations of the uninterrupted increases in Vostok temperatures from 7962 BC to 1879 (TDv). In addition, the maximum duration of the increases in global temperatures equals 37 years, while the mean duration of the increases in the Vostok temperatures equals 99 years. [50]

Conclusion 13—The results reject the hypothesis that the maximum duration of the uninterrupted increases in global temperature from 1880 to 2014 is significantly longer than the durations of the uninterrupted increases in Vostok temperatures from 7962 BC to 1879. The maximum duration of the increases in global temperature is even shorter than the natural average.

Conclusion 14—The magnitude (Conclusion 11), slope (Conclusion 12), and duration (Conclusion 13) of the recent increase in the average global temperature have failed to reach statistical significance. Hence, that thermal increase is random, rather than systematic. This suggests that the average global temperature is not the driving force behind the symptoms of climate change. Instead, a missing variable seems to be driving both the average global temperature and the symptoms of climate change. Hence, the symptoms of climate change such as the melting ice-sheets were observed without a significant increase in the average global temperature. There was climate change before the global average temperature was significant. Consequently, even a significant global average temperature in the future would be insufficient evidence for the notion that it is the only driving force behind climate change.

Thermal gap

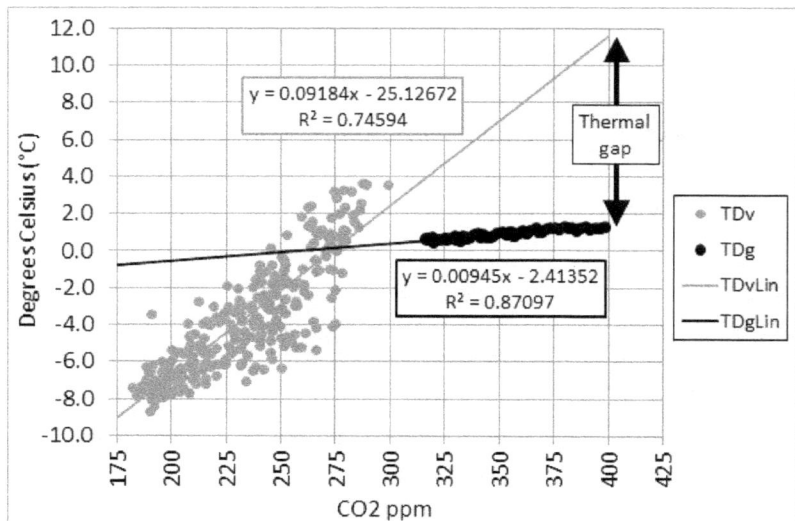

Figure 7—Superimposed, Figure 2 and Figure 5 show the thermal gap of 10.1 °C at the current CO2 of 399 ppm.[51] TDv = Vostok temperature as a function of CO2 (Figure 2); and TDg = global annual temperature as a function of Hawaiian CO2 (Figure 5).[52] TDvLin (Equation 1) and TDgLin (Equation 3) are adequate linear functions of TDv and TDg, because they explain respectively 75% and 87% of the variances.

Aim 12—Comparing the thermal predictions of the Vostok temperatures as a linear function of the Vostok CO2 from 412,096 to 9,024 BC (TDvLin Equation 1) with the thermal predictions of the global temperatures as a linear function of the Hawaiian CO2 from 1959 to 2014 (TDgLin Equation 3).[53]

Result 12—Figure 7 shows that TDvLin (0.09184°C/ppm) is about 10 times steeper than TDgLin (0.00945°C/ppm). In accord, at the CO2 of 399 ppm in 2014, TDvLin predicts a temperature of 11.5 °C, while TDgLin predicts 1.4 °C.[54] Hence, the thermal gap at a CO_2 of 399 ppm equals: 11.5 − 1.4 = 10.1 °C in 2014.[55] This assumes that Equation 1 represents the long-term relationship between temperature and CO_2. It is proposed that natural glacial cooling fosters the thermal gap.

Result 13—At a CO_2 of 0 ppm, TDvLin predicts that the temperature would be -25.1 °C, while TDgLin predicts -2.4 °C.[56] Hence, the thermal gap at a CO_2 of 0 ppm equals: (-2.4) - (- 25.1) = 22.7 °C.[57]

Result 14—The estimated lowest possible natural glacial temperature would be the global temperature in 2014 minus the thermal gap in 2014, which equals: 1.4 − 10.1 = -8.7 °C in 2014.[58] If the glacial period has already started, then the real natural glacial temperature is masked by the positive thermal effect of artificial CO_2. Hence, the thermal gap can only provide information about the lowest possible natural temperature. The true natural temperature in 2014 is hidden and cannot be derived from the available datasets.

Conclusion 15—Equation 1 and Equation 3 predict a thermal gap in 2014 of 10.1 °C at a CO_2 of 399 ppm. On the other hand, those equations predict a thermal gap of 22.7 °C at a CO_2 of 0. Furthermore, those equations estimate a lowest possible natural glacial temperature of -8.7 °C in 2014.

Aim 13—Testing the hypothesis that the thermal gap in 2014 of 10.1 °C at a CO2 of 399 ppm is statistically significant.

Result 15—A randomized probability test of the thermal gap in 2014 of 10.1 °C at a CO2 of 399 ppm yielded the following statistics: p = 0.0000; i = 10,000.[59] The result shows that p < 0.05 and, therefore, it supports Aim 13—Testing the hypothesis that the thermal gap in 2014 of 10.1 °C at a CO2 of 399 ppm is statistically significant. In addition, the unexplained thermal gap of 10.1 °C is rather large in comparison to the alleged global warming of 1.4 °C.[60]

Conclusion 16—The results support the notion that the thermal gap of 10.1 °C is statistically significant. This suggests that a missing variable has changed recently the relationship between temperature and CO2.

Conclusion 17—The results support the notion that the current average global temperature of 1.4 °C is 10.1 °C lower than predicted by 420 Kyrs of Vostok data. To explain this thermal gap, it is proposed that the interglacial period ended some time ago and that we live in a natural glacial period with a temperature of -8.7 °C. Artificial CO2 would increase that glacial temperature with 10.1 °C to the observed global temperature of 1.4 °C in 2014.

Notes

SUMMARY ANALYSES

Conclusion 18—The Hawaiian CO_2 of 399 ppm in 2014, and the significant correlations between CO_2 and temperature, provide non-conclusive support for both the unilateral and bilateral hypotheses of climate change (Conclusion 1, Conclusion 6, and Conclusion 10). Nevertheless, the unilateral hypothesis of climate change has to be rejected, because the magnitude, slope, and duration of the recent increases in global temperature failed to reach statistical significance (Conclusion 11, Conclusion 12, Conclusion 13, and Conclusion 14). In addition, the global temperature in 2014 of 1.4 °C is significantly lower than predicted by Vostok CO_2 (Conclusion 15, and Conclusion 16). In contrast to the unilateral hypothesis, the bilateral hypothesis of climate change is supported by the significant thermal stability of the current interglacial (Conclusion 4); the significant duration of the current interglacial (Conclusion 5); and the significant thermal gap of 10.1 °C (Conclusion 2, Conclusion 7, Conclusion 8, Conclusion 15, and Conclusion 16). In addition, this study supports the notion that the glacial period started before the industrial revolution (Conclusion 4, Conclusion 5, and Conclusion 17). It also rejects the plans of the scientific-political establishment to stabilize CO_2 at 400 ppm, because that is unlikely to reduce the global temperature to normal (Conclusion 3 and Conclusion 9).

DISCUSSION

This pilot study differentiates between unilateral and bilateral climate change. The widely accepted unilateral hypothesis of climate change (global warming) holds that artificial greenhouse gasses, such as carbon dioxide (CO_2), increase the global temperature above normal by trapping heat in the atmosphere. In contrast, the proposed bilateral hypothesis of climate change holds that the artificial global warming compensates for the natural global cooling that is caused by the start of a glacial period. The bilateral hypothesis implies that the climate is destabilized by the opposing forces of artificial global warming and natural global cooling. The bilateral hypothesis extends, rather than rejects the unilateral hypothesis.

Within the context of the unilateral hypothesis, global warming is defined as the average global temperature in reference to the Vostok interglacial baseline of 0.0 °C. Within that context, the artificial global warming would be the global average temperature of 1.4 °C as observed in 2014.

Within the context of the bilateral hypothesis, global warming is defined as the average global temperature in reference to the natural global temperature. The Vostok CO_2 predicts a temperature of 11.5 °C in 2014, while the observed average global temperature equals at the same time only 1.4 °C. This supports the notion that the natural global cooling equals 10.1 °C.

Therefore, the estimated natural temperature equals -8.7 °C in 2014. This natural temperature would be increased by the artificial global warming of 10.1 °C to the average global temperature of 1.4 °C as observed in 2014. Hence, within the context of the bilateral hypothesis, global warming would be 10.1 °C, rather than 1.4 °C.

The statistical analyses in this study suggest that the average unilateral global temperature of 1.4 °C is within the natural interglacial limits. Consequently, it has to be considered a random effect, rather than a systematic one. On its own, such a random effect cannot account for the recent instability of the climate. Nevertheless, it might aggravate the symptoms of that instability.

The bilateral hypothesis of climate change is presented in order to explain the recent instability of the climate without having to rely on the non-significant increase in the average global temperature of 1.4 °C. It is proposed that the opposing forces of artificial global warming and natural glacial cooling create a climatological pendulum. The swings of that pendulum would drive the recent instability of the climate and, therefore, it would melt the polar ice and cause rising sea levels.

Within the context of the unilateral hypothesis, the decision of the scientific-political establishment to stabilize the CO_2 at about 400 ppm is likely to be inadequate. The results of this study suggest that the CO_2 might have to be lowered to 255 ppm in order to return to the normal interglacial temperature of 0.0 °C. Even at that normal temperature, the precarious balance of the opposing forces of bilateral climate change might maintain the recent instability of the climate.

Within the context of the bilateral hypothesis, the artificial CO2 is needed to maintain an interglacial temperature of about 0.0 °C during the current glacial period by balancing artificial global warming and natural global cooling. Hence, CO2 would be a beneficial, rather than a hazardous substance. It is questionable whether humanity is able to produce enough artificial CO2 to maintain that interglacial temperature for the next 90,000 years of the glacial period.

The results of this pilot study have significant implications for the definition, measurement, and management of climate change. Furthermore, it justifies an urgent study of the bilateral hypothesis of climate change. This hypothesis implies that the scientific-political establishment might have to manage actively the atmospheric CO2 concentration, rather than stabilizing it. Nevertheless, this study provides more questions than answers, because the fundamental weakness of climatology is the lack of valid and reliable long-term global datasets.

Notes

APPENDICES

BIBLIOGRAPHY

GISTEMP Team, N. G. (2015). GISS Surface Temperature Analysis (GISTEMP). *Dataset accessed 2015-05-03 at http://data.giss.nasa.gov/gistemp/.*

Gore, A. (2006). *An Inconvenient Truth.* Amazon.

Hansen, J. R. (2010). Global surface temperature change. *Rev. Geophys., 48, RG4004, doi:10.1029/2010RG000345.*

Mauna Loa Observatory (Scripps / NOAA / ESRL). (2015). *Monthly Mean CO2 Concentrations (ppm).* http://co2now.org/Current-CO2/CO2-Now/noaa-mauna-loa-co2-data.html.

Moore, G. P., & McCabe, D. S. (2003). *Introduction to the Practice of Statistics.* New York: W. H. Freeman and Company.

NOAA/NGDC Paleoclimatology Program, B. C. (n.d.). Vostok Ice Core Data for 420,000 Years.

Petit, J. R. (2001). *Vostok Ice Core Data for 420,000 Years.* Boulder CO, USA: IGBP PAGES/World Data Center for Paleoclimatology Data Contribution Series #2001-076. NOAA/NGDC Paleoclimatology Program.

Petit, J. R., Jouzel, J., Raynaud, D., Barkov, N. I., M, B. J., Basile, I., et al. (1999). Climate and Atmospheric History of the Past 420,000 years from the Vostok Ice Core, Antarctica. *Nature, 399,* 429-436.

Schade, A. (2015). bioPAD Nemonik Thinking (PowerPoint). nemonik-thinking.org.

Schade, A. (2016). *Global Warming is the Solution.* nemonik-thinking.org.

Schade, A. (2016). *Glossary Nemonik Thinking.* nemonik-thinking.org.

Schade, A. (2016). *Lao Zi's Dao De Jing for Nemonik Thinkers.* nemonik-thinking.org.

Schade, A. (2016). *Think Smarter with Nemonik Thinking.* nemonik-thinking.org.

Schade, A. (planned 2017). *Dictionary Nemonik Thinking.* nemonik-thinking.org.

Schade, A. (planned 2017). *Sun Zi's Strategies for Nemonik Thinkers.* nemonik-thinking.org.

GLOSSARY

Alpha (α)—probability that a decision maker accepts to make a mistake. In statistics, an α-value would range traditionally from lax (0.05) to strict (0.001).

Alternative hypothesis (Ha)—statement that the effect is systematic, rather than random. See Hypothesis and Null hypothesis (Ho).

Antithesis—description of reality that contradicts a thesis. See Hegel.

Artificial global warming—comprises domestic and industrial global warming.

Artificial—manmade.

Baseline—the thermal baseline, or normal temperature, is the average atmospheric temperature in Vostok from calendar year 989 to 1989 set at 0.00 °C (Vostok Kyr1 BP = 0.00 °C).

Bilateral-climate change-hypothesis—holds that the effect of artificial global warming compensates for the effect of natural global cooling. Low glacial temperatures are increased by solar heat that is trapped by domestic and industrial atmospheric greenhouse gasses such as carbon dioxide (CO_2).

BP—Before Present. For the Vostok dataset of 420 Kyrs that is calendar year 1989.

Carbon dioxide (CO2)—greenhouse gas that allegedly traps heat in the atmosphere.

Climate change—long-term and fundamental change in the global climate. Climate change comprises long-term natural cycles of global warming and cooling, and recently artificial domestic and industrial global warming.

Climate control—voluntary and involuntary artificial ways to control the natural climate.

Climatology—study of the climate.

CO2—See carbon dioxide.

Confounding variables—variables not included in the statistical analyses and are not eliminated by experimental design, which disturb the results.

Criterion—See Alpha.

Data point—numerical single observation, fact, measurement, or score in a dataset. For example: 1.0 is a data point in the dataset (2.0, 3.0, -2.0, **1.0**, and -1.0).

Data snooping—selecting a part of the dataset under examination in order to find a significant result. Data snooping breaches the process of random selection in which each data point has an equal likelihood to be selected. Consequently, the selected part might not represent the population and could cause spurious results. Data snooping is used as an exploratory tool to create new hypotheses.

Dataset—collection of data points. For example (2.0, 3.0, -2.0, 1.0, and -1.0).

Descriptive statistics—branch of statistics concerned with computing numbers (statistics) that describe the centricity and spread of data points comprising a dataset. Descriptive statistics include the mean $(\bar{x})$, standard deviation (STD), maximum (max), minimum (min), range, and the number of data points (n).

Domestic global warming—hypothesis that humanity has unknowingly affected the climate by releasing domestic greenhouse gasses. Domestic greenhouse gasses are likely to have increased with decreasing atmospheric temperatures and increasing global population. See Global warming.

Facts—testable descriptions of reality that are supported adequately by sensory perception and rational thinking. See Hypothesis.

Glacial decline—thermal decline at the onset of a glacial period. It stretches from the interglacial thermal peak to an arbitrary limit of -4.00 degrees Celsius (°C) relative to the Vostok base Kyr1 BP = 0.00 °C.

Glacial period—long-term climatological period with mean millennial temperature deviations below the mean temperature deviation of the Vostok base Kyr1 BP = 0.00 °C.

Global cooling—long-term and fundamental decrease in the global temperature. Global cooling comprises natural and artificial decreases in global temperature. See Climate change.

Global warming—long-term and fundamental increase in the global temperature. Within the context of the unilateral hypothesis, global warming is defined as the average global temperature in reference to the Vostok interglacial baseline of 0.0 °C. Within the context of the bilateral hypothesis, global warming is defined as the average global temperature in reference to the natural global temperature of -8.7 °C in 2014. See Climate change.

Greenhouse gasses—gasses trapping heat in the atmosphere. See carbon dioxide.

Ha—See Alternative hypothesis.

Hegel, Frederich (1770-1831)—German philosopher who introduced a dialectic that describes the progress of knowledge. The dialectic holds that each thesis evokes an antithesis, which is followed by a synthesis. This synthesis becomes the new thesis. Hegel's dialectic is a process of successive approximation.

Ho—See Null hypothesis.

Hypothesis—untested, but testable description of reality. See Null hypothesis (Ho), Alternative hypothesis (Ha), and Facts.

Industrial global warming—hypothesis that humanity's industry has pollutes the climate by releasing too much greenhouse gasses. See Climate change.

Interglacial period—long-term climatological period with mean temperature deviations per Kyr equal or above the mean temperature deviation of the Vostok base Kyr1 BP = 0.00 °C. The mean duration of past interglacial periods is about 4.6 Kyrs.

Kyr—millennium or period of 1,000 years.

Maximum (max)—data point with the largest magnitude in a dataset. For the dataset (2.0, **3.0**, -2.0, 1.0, and -1.0), the maximum = 3.0.

Mean or average ($\bar{x}$)—statistic that measures the central value of a dataset, which is the average of the data points. The mean is computed with the equation: $\bar{x} = \frac{(x_1 + x_2 + \cdots x_n)}{n}$. In which, $x_1, x_2, etc.$ are data points, and (n) is the number of data points. For the dataset (2.0, 3.0, -2.0, 1.0, and -1.0), the mean equals: (2.0 + 3.0 + -2.0 + 1.0 + -1.0) / 5 = 3.0 / 5 = 0.6.

Median—statistic that measures the central value of a dataset, which is the number in the middle of that dataset sorted by magnitude. For example, the sorted dataset (2.0, 3.0, -2.0, 1.0, and -1.0) would be (3.0, 2.0, **1.0**, -1.0, -2.0) and the number in the middle = 1.0.

Millennium—1000 years or Kyr.

Minimum (min)—data point with the smallest magnitude in a dataset. For the dataset (2.0, 3.0, **-2.0**, 1.0, and -1.0), the minimum = -2.0.

Missing variable—variable that is not included in the analysis.

Natural global cooling—part of the natural climatological cycle characterized by long-term decreasing temperatures. See Global warming.

Natural global warming—part of the natural climatological cycle characterized by long-term increasing temperatures. See Global warming.

Natural—not manmade. Opposite of artificial.

Nemonik thinking—exhaustive cognitive system that synthesizes rational, creative, and intuitive thinking. It maximizes success by evaluating 17 nemoniks, which are memorized keywords describing all the perceived aspects of the mind, reality, and their interaction. As defined by Lao Tzu—"Success is obtaining what you seek and escaping what you suffer."

Nemonik-accelerator—increases the speed of thinking by fostering agreement during disagreement, while fostering disagreement during agreement. This nemonik tool is based on Hegel's dialectic.

n—number of data points in a dataset. For the dataset (2.0, 3.0, -2.0, 1.0, and -1.0), n = 5.

Normal or Gaussian frequency distribution—dataset described by a symmetric, unimodal, and bell shaped frequency curve. The mean and the median of a normal frequency distribution are the same. In parametric statistics, a normal population distribution is a basis for evaluating of data points and samples.

Normal temperature—See Baseline.

Null hypothesis (Ho)—statement that the effect is random, rather than systematic. Null refers to the lack of a systematic effect. See Hypothesis and Alternative hypothesis (Ha).

One-tailed p-test—evaluates the probability (p) that a data point is significant higher than the mean, or alternatively, that the data point is significant lower than the mean. For example, the evaluation of global warming (only warming) would require a one-tailed p-test, while climate change (warming and cooling) requires a two-tailed p-test. The criterion for a one-tailed test equals (α). Hence, the unilateral and bilateral hypotheses require different tests. See Two-tailed p-test.

Parameters—numbers describing the population distribution. Parameters include the mean ($\bar{x}$), standard deviation (STD), maximum (max), minimum (min), range, and the number of data points (n). See Statistics (descriptive).

Parametric statistics—branch of statistics concerned with evaluating the probability that an effect is either systematic or random, by comparing the statistics of a sample with the parameters of a normal dataset. For that reason, the probabilities of making incorrect decisions (Type I and II errors) are computed. The main weakness of parametric statistics is the assumption that the data point or sample under examination is part of a population with a normal frequency distribution.

ppm—unit to measure gas meaning Parts Per Million. Alternatively, ppm is referred to as ppmv.

ppmv—parts per million volume. See ppm.

Probability (p)—computed probability of making a Type I error. See Alpha.

p—See Probability.

Random effect—effect attributed to random variation around the mean. Opposite: Systematic effect.

Randomization test—non-parametric test that computes the significance of statistics by comparing the original statistics of the sample with iterated statistics obtained from the randomized sample. This test creates its own population distribution, which is based on the data points of the sample. The randomization test is one of the most versatile tests in statistics. It can be used to test the significance of means, medians, correlation coefficients, linear functions, etc.

Range—is the maximum of the data points minus the minimum of the data points. For the dataset (2.0, **3.0**, **-2.0**, 1.0, and -1.0), the range = 3.0 – -2.0 = 3.0 + 2.0 = 5.0.

Regression towards the mean—if the size of a sample is increased and the mean of that larger sample moves towards the mean of the population. Therefore, extreme high means are likely to decrease, while extreme low means are likely to increase with an increase of the sample size.

Significant—computed probability of making an error (p) is smaller than the accepted probability of making an error (α). A significant effect is considered systematic, rather than random.

Standard deviation (STD)—measures the spread or variation of the data points around their mean. The STD is computed with the equation: $STD = \sqrt{(\sum(x_i - \bar{x})^2) / (n - 1)}$. In which, $\sqrt{}$ = square root; $\sum$ = sum or total; x_i = all data points in the dataset from x_i to x_n; $\bar{x}$ = mean of the dataset; n = number of data points in the dataset. For dataset (2.0,

3.0, -2.0, 1.0, and -1.0); the mean = 0.6; and n = 5. In accord, STD $= \sqrt{(\sum((2.0 - 0.6)^2 + (3.0 - 0.6)^2 + (-2.0 - 0.6)^2 + (1.0 - 0.6)^2 + (-1.0 - 0.6)^2) / (5 -1))}$ $= \sqrt{(\sum((1.4)^2 + (2.4)^2 + (-2.6)^2 + (0.4)^2 + (-1.6)^2) / 4)}$ $= \sqrt{(\sum(1.96 + 5.76 + 6.76 + 0.16 + 2.56) / 4)} = \sqrt{(17.20 / 4)} = \sqrt{4.30} = 2.1$.

Standard-score—See z-score.

Statistic (descriptive)—number that describes the centricity, spread, or size of a dataset sampled from a population.

Statistics—field of study concerned with collecting, organizing, describing, analysing, and evaluating numerical facts called data.

Synthesis—merger of separate parts of a phenomenon into an entirety. In the context of Hegel's dialectic, a synthesis is a description of reality that merges a thesis and an antithesis into a new thesis. A synthesis might create synergy. See Hegel.

Systematic effect—effect attributed to a specific variable. Opposite Random effect.

Temperature deviations—temperatures expressed as differences from a mean temperature during a base period. Deviations are also indicated by the Greek letter delta. Temperature deviations could be expressed as degrees Celsius, Fahrenheit, or Kelvin. See Absolute temperatures.

Thermal baseline—See Baseline.

Thermal gap—difference between the estimated and actual annual global atmospheric temperature deviations.

Thermal outliers—temperature deviations that are both higher than the global maximum of 1.32 °C in 2014 and associated with CO_2 concentrations below the Hawaiian maximum of 398.6 ppm in 2014.

Thermal—refers to temperature.

Thesis—a description of reality. See Hegel.

Trend—a one directional change in the data across time. See also Cycles.

Two-tailed p-test—evaluates the probability (p) that a data point is either significant higher or lower than the mean. For example, the evaluation of global warming (only warming) would require a one-tailed p-test, while climate change (warming and cooling) requires a two-tailed p-test. The criterion for a two-tailed test equals α / 2. See One-tailed p-test.

Type I error—incorrect decision to reject a true null hypothesis (Ho) and, therefore, accept the false alternative hypothesis

(Ha). This error leads to the incorrect conclusion that a random effect is systematic. Opposite Type II error.

Type II error—incorrect decision to accept a false null hypothesis (Ho) and, therefore, reject the alternative hypothesis (Ha). This error leads to the incorrect conclusion that a systematic effect is random. Opposite Type I error.

Unilateral-climate change-hypothesis—holds that industrial greenhouse gasses, such as carbon dioxide (CO_2), increase the global temperature by trapping heat in the atmosphere.

z-score or standard-score—difference between a data point and the mean, expressed in standard deviations. The z-score is computed with the equation: $z = (x - \bar{x}) / STD$. In which, x = the data point under examination; $\bar{x}$ = mean of the dataset; and STD = standard deviation of the dataset. For dataset (2.0, **3.0**, -2.0, 1.0, and -1.0), $\bar{x}$ = 0.6; STD = 2.07; and the data point to be tested = 3.0. The z-score of 3.0 = (3.0 – 0.6) / 2.07 = 2.4 / 2.07 = 1.16. Based on tables representing a normal distribution, the p-value for the z-score of 1.16 = 0.1230 (Moore & McCabe, 2003). Hence, p 0.1230 > α (0.05) and, therefore, the data point (3.0) is not significantly higher than the mean of the dataset (0.6). The difference between that data point and the mean has to be attributed to a random effect. In contrast, a data point of 6.0 would obtain a z-score = (6.0 – 0.6) / 2.07 = 5.4 / 2.07 =

2.60. The p-value for the z-score of 2.60 = 0.0047. Hence, p 0.0047 < α (0.05) and, therefore, the data point (6.0) is significantly higher than the mean of the dataset (0.6). The difference between that data point and the mean has to be attributed to a systematic effect.

α (alpha)—See Alpha.

LISTS

Aims

Figures

Equations

Equation 1 (Figure 2)—Describes the Vostok temperature TDv as a linear function of the Vostok CO2v (TDvLin): $y = 0.09184x - 25.12672$. In which: $y = $ TDv and $x = $ CO2v. The linear function (TDvLin) and the second order polynomial function (TDvPol) shown in Figure 2 are both adequate, because they explain respectively 75% and 76% of the variance in TDv. Increasing the sensitivity of the polynomial function to the sixth order did not improve significantly the explained variance of TDvPol. The linear function was preferred, because it provides conservative predictions, optimizes the simplicity, and facilitates the comparison between the global and Vostok results................20

Equation 2—The thermal stability index (TSI) is introduced, which equals: TSI = 1 / STD. In which, STD is the standard deviation of the Vostok detrended millennial temperatures (TDv) within a window of 10 Kyrs. STD is a measure that changes in synchrony with the thermal instability. In contrast, TSI is a measure that changes in synchrony with the thermal stability. To determine the TSI values within an observation period, the window slides across that period in steps of 1 Kyr. 23

Equation 3 (Figure 5)—Describes the global temperature TDg as a linear function of Hawaiian CO2h (TDgLin): $y = $

Results

Result 1—A randomized probability test of the ranked correlations between TDv and CO2v, from 414,085 to 2,342 BP, yielded: $r = 0.87$; and $p = 0.0000$. The result shows that p < 0.05 and, therefore, it supports Aim 2—Testing the hypothesis that the correlation between the Vostok temperatures (TDv) and the Vostok CO2 (CO2v) during the previous 420 Kyrs is statistically significant. Nevertheless, during the period from 3,634 to 2,342 BP, the CO2v increased with 12 ppm, while TDv decreased with 1.7 °C. This suggests that one or more missing variables are involved. A missing variable is a variable that is not included in the analysis. 18

Result 2—Equation 1 predicts a temperature of 11.6 °C above the baseline at the current CO2 of 399 ppm. On the other hand, Equation 1 predicts a temperature of -25.1 °C below the baseline at a CO2 of 0 ppm. Furthermore, Equation 1 predicts that the current CO2 of 399 ppm has to be decreased with 125 ppm to 274 ppm in order to return the temperature to its baseline..21

Result 3—The mean TSI values obtained with sliding windows of 10 Kyrs yielded for the observation period from Kyr423 to Kyr11 BP the following statistics: minimum = 0.26; maximum = 4.02; mean = 1.33; STD = 0.72; and n = 404. In

Result 5—A randomized probability test of the correlation between the ranked TDg and CO2h values from 1959 to 2014 yielded: r = 0.92 and p = 0.0000. The result shows that p < 0.05 and, therefore, it supports Aim 6—Testing the hypothesis that the correlation between the average global temperatures (TDg) and the Hawaiian CO2 (CO2h) from 1959 to 2014 is statistically significant...27

Result 6—Equation 3 predicts a temperature of 1.4 °C above the baseline at the current CO2 of 399 ppm. On the other hand, Equation 3 predicts a temperature of -2.4 °C below the baseline at a CO2 of 0 ppm. Furthermore, Equation 3 predicts that the current CO2 of 399 ppm has to be decreased with 144 ppm to 255 ppm in order to return the temperature to its baseline...28

Result 7—Visual inspection of the CO2 in [Figure 1—Vostok temperature (TDv) and Vostok CO2 (CO2v) from 414,085 to 2,342 BP with irregular data intervals.] and [Figure 4—Global annual temperature (TDg) and Hawaiian annual CO2 (CO2h) from calendar year 1959 to 2014.] seems to suggest that the current Hawaiian CO2 of 399 ppm is at an all–time high. However, the data interval for the Vostok maximum CO2v of 299 ppm equals 681 years, while the data interval for the maximum CO2h of 399 ppm equals only one year. Therefore, regression towards the mean is likely to confound the comparison. For example, the mean of the CO2h from 1959 to 2014 equals 351 ppm. Hence, increasing the data interval from 1 to 55 years decreases the mean CO2h from 399 to 351 ppm. Despite this large decrease, the Hawaiian data interval of 55 years is still much shorter than the Vostok data interval of 681 years. At this stage, one can only speculate about the magnitude of the annual Vostok CO2 in the past. No one really knows. Therefore, regression towards the mean is still likely to confound the comparison between the short-term CO2h and the long-term CO2v. 30

Result 8—Figure 6 shows that the maximum global window temperature in 2014 (TDgSWm = 0.9 °C) is higher than the mean Vostok window temperature from 7962 BC to 1928 (TDvSWx = 0.1 °C). However, during five separate occasions, the Vostok TDvSW values have been higher than the global thermal maximum (TDgSWm = 0.9 °C). In addition, the Vostok thermal maximum (TDvSWm = 2.3 °C) is considerably higher than the global thermal maximum (TDgSWm = 0.9 °C). 33

Result 9—If the temperature is below its baseline, then there is by definition no global warming. Hence, the evaluation of global warming makes only sense if the temperature is above the baseline. In accord, during 4,671 of the 9,891 observation years the Vostok interglacial window temperatures (TDvSW) were higher than the baseline. In addition, during 518 of those 4,671 decision years, the TDvSW values exceeded the global thermal maximum (TDgSWm = 0.9 °C). Hence, p = 518 / 4671 = 0.11. The result shows that p > 0.05 and, therefore, it rejects Aim 9—Testing the hypothesis that the global average temperature of **1.4** °C in 2014 (TDg) is significantly higher than the Vostok interglacial temperatures from 7962 BC to 1928 (TDv)..33

Result 10—The slopes of 26 of the 51 increases in the Vostok interglacial window temperatures were steeper than the maximum slope of the global window temperatures of 0.00858 °C/year. Hence, the probability equals: p = 26 /51 = 0.51. The result shows that p > 0.05 and, therefore, it rejects Aim 10—Testing the hypothesis that the maximum slope of the uninterrupted increases in global temperatures from 1880 to 2014 (TDg) is significantly steeper than the slopes of the interrupted increases in the Vostok interglacial temperatures from 7962 BC to 1879 (TDv). In addition, the mean slope of the increases in Vostok temperatures of 0.01017 °C/year is steeper than the maximum slope of the increases in global temperatures of 0.00858 °C/year..35

Result 11—Of the 51 durations of the uninterrupted increases in the Vostok temperatures, 50 lasted longer than the maximum duration of the uninterrupted increases in the global temperature of 37 years. Hence, the probability equals: p = 50 / 51 = 0.98. The result shows that p > 0.05 and, therefore, it rejects Aim 11—Testing the hypothesis that the maximum duration of the uninterrupted increases in global temperature from 1880 to 2014 (TDg) is significantly longer than the durations of the uninterrupted increases in Vostok temperatures from 7962 BC to 1879 (TDv). In addition, the maximum duration of the increases in global temperatures equals 37 years, while the mean duration of the increases in the Vostok temperatures equals 99 years. 36

Result 12—Figure 7 shows that TDvLin (0.09184°C/ppm) is about 10 times steeper than TDgLin (0.00945°C/ppm). In accord, at the CO_2 of 399 ppm in 2014, TDvLin predicts a temperature of 11.5 °C, while TDgLin predicts 1.4 °C. Hence, the thermal gap at a CO_2 of 399 ppm equals: 11.5 − 1.4 = 10.1 °C in 2014. This assumes that Equation 1 represents the long-term relationship between temperature and CO_2. It is proposed that natural glacial cooling fosters the thermal gap... 38

Result 13—At a CO_2 of 0 ppm, TDvLin predicts that the temperature would be -25.1 °C, while TDgLin predicts -2.4 °C. Hence, the thermal gap at a CO_2 of 0 ppm equals: (-2.4) - (- 25.1) = 22.7 °C.. 39

Result 14—The estimated lowest possible natural glacial temperature would be the global temperature in 2014 minus the thermal gap in 2014, which equals: $1.4 - 10.1 = -8.7$ °C in 2014. If the glacial period has already started, then the real natural glacial temperature is masked by the positive thermal effect of artificial CO_2. Hence, the thermal gap can only provide information about the lowest possible natural temperature. The true natural temperature in 2014 is hidden and cannot be derived from the available datasets..................39

Result 15—A randomized probability test of the thermal gap in 2014 of 10.1 °C at a CO2 of 399 ppm yielded the following statistics: $p = 0.0000$; $i = 10,000$. The result shows that $p < 0.05$ and, therefore, it supports Aim 13—Testing the hypothesis that the thermal gap in 2014 of 10.1 °C at a CO2 of 399 ppm is statistically significant. In addition, the unexplained thermal gap of 10.1 °C is rather large in comparison to the alleged global warming of 1.4 °C..............40

Conclusions

CO2 of 0. Furthermore, those equations estimate a lowest possible natural glacial temperature of -8.7 °C in 2014. 39

Conclusion 16—The results support the notion that the thermal gap of 10.1 °C is statistically significant. This suggests that a missing variable has changed recently the relationship between temperature and CO2. 40

Conclusion 17—The results support the notion that the current average global temperature of 1.4 °C is 10.1 °C lower than predicted by 420 Kyrs of Vostok data. To explain this thermal gap, it is proposed that the interglacial period ended some time ago and that we live in a natural glacial period with a temperature of -8.7 °C. Artificial CO2 would increase that glacial temperature with 10.1 °C to the observed global temperature of 1.4 °C in 2014. 40

Conclusion 18—The Hawaiian CO2 of 399 ppm in 2014, and the significant correlations between CO2 and temperature, provide non-conclusive support for both the unilateral and bilateral hypotheses of climate change (Conclusion 1, Conclusion 6, and Conclusion 10). Nevertheless, the unilateral hypothesis of climate change has to be rejected, because the magnitude, slope, and duration of the recent increases in global temperature failed to reach statistical significance (Conclusion 11, Conclusion 12, Conclusion 13, and Conclusion 14). In addition, the global temperature in

2014 of 1.4 °C is significantly lower than predicted by Vostok CO_2 (Conclusion 15, and Conclusion 16). In contrast to the unilateral hypothesis, the bilateral hypothesis of climate change is supported by the significant thermal stability of the current interglacial (Conclusion 4); the significant duration of the current interglacial (Conclusion 5); and the significant thermal gap of 10.1 °C (Conclusion 2, Conclusion 7, Conclusion 8, Conclusion 15, and Conclusion 16). In addition, this study supports the notion that the glacial period started before the industrial revolution (Conclusion 4, Conclusion 5, and Conclusion 17). It also rejects the plans of the scientific-political establishment to stabilize CO_2 at 400 ppm, because that is unlikely to reduce the global temperature to normal (Conclusion 3 and Conclusion 9)............................ 42

MY OTHER BOOKS

Global Warming

Global Warming is the Solution (Schade, 2016).

This book presents a bilateral hypothesis for climate change. Mainstream climatology lacks scientific integrity and statistical methodology. Peer review is changed into peer pressure and objectors are silenced by labelling them '*Deniers*'. Proper statistical analyses are replaced by fancy graphs and non-causal correlation analyses. The conclusions are predominantly based on the previous 166 years, while more than 420,000 years of Antarctic data are ignored. Climatology also ignores the solar expert Professor Zharkova, who predicts a mini ice-age by 2030. The present study shows that the current 399 ppm of CO_2 predicts a global temperature of 11.5 °C. It also shows that the observed global temperature of 1.3 °C failed to reach statistical significance. In addition, the data support the hypothesis that we live in a glacial period. This hypothesis is supported by the thermal gap of CO_2, the long interglacial duration, and the interglacial thermal stability. Consequently, decreasing atmospheric CO_2 could induce glacial conditions threatening the survival of humanity.

Download a free eBook version

@ nemonik-thinking.org

Nemonik Thinking

Think Smarter with Nemonik Thinking (Schade, 2016).

This is the operating manual for your mind that you should have received at birth. Nemonik thinking is a smarter way of thinking that aims to maximize your success by evaluating seventeen nemoniks, which are memorized keywords describing all the perceived aspects of your mind, reality, and their interaction. Success is obtaining what you seek and escaping what you suffer. To maximize that success, nemonik thinking mobilizes the hidden genius, accelerates thinking, improves memory, reveals opportunities and threats, creates questions and ideas, and reduces stress levels. It is like playing a musical keyboard with seventeen keys producing an infinite repertoire of smart strategies. Nemonik thinking is unique because it is the first exhaustive and transferable way of thinking. In contrast, conventional thinking is time consuming. Hence, the less time you have, the greater the necessity to study nemonik thinking. You might be the smartest thinker in the world, but only nemonik thinking could make you the smartest thinker you can be.

Download a free eBook version
@ nemonik-thinking.org

Nemonik Glossary

Glossary of Nemonik Thinking (Schade, 2016).

Nemonik thinking is a competitive advantage because it mobilizes the hidden genius, accelerates thinking, improves memory, prevents blind-spots, and reveals opportunities, while its constant preparedness reduces stress levels. Definitions, associated with the mind and reality, are inherently hypothetical, fuzzy, and intertwined. Nevertheless, to improve our understanding of the way we think, we have to identify, differentiate, and define those components. Therefore, this glossary provides descriptions for the concepts associated with nemonik thinking. To become skilled in nemonik thinking, it is recommended to study—*Think Smarter with Nemonik Thinking (Schade, 2016).*

Download a free eBook version
@ nemonik-thinking.org

Nemonik Dictionary

Dictionary Nemonik Thinking (Schade 2016).

Nemonik thinking mobilizes the hidden genius, accelerates thinking, improves memory, reveals opportunities and threats, creates questions and ideas, and reduces stress levels. Nemonik thinking divides the mind into 17 nemonik regions. That division defragments information, which facilitates the storage, maintenance, recall, and processing of associated information from memory. However, the boundaries of the nemonik regions are fuzzy. Therefore, the aim of this dictionary is to differentiate them by providing keywords for each nemonik concept. The first part of this dictionary translates nemonik concepts into common keywords e.g. *advance* into attack, bypass, etc. In contrast, the second part translates common keywords into nemonik concepts e.g. attack, bypass, etc. into *advance*. This dictionary shows that the complexity of conventional thinking comprises thousands of keywords that can be simplified to 17 nemoniks. This reduction will increase the speed of your thinking. To become skilled in nemonik thinking, it is recommended to study—*Think Smarter with Nemonik Thinking (Schade, 2016).*

Download a free eBook version

@ nemonik-thinking.org

Education Kills Humanity

Education Kills Humanity (Schade, 2016).

Humanity is facing huge manmade problems such as over-population, dwindling resources, pollution, climate change, and warfare. Nevertheless, we should not blame corrupt politicians, uncaring industrialists, greedy investors, passionate greenies, and warmongers. They are the products of our educational system, which conditions students with ratings to maximize the probability of winning. Winning is defeating opponents in competition. Therefore, conventional thinking is conflict oriented, which fosters aggression, control, effort, and force. This inhibits the truth and, therefore, it is self-destructive. The educational failure is maintained by cognitive dissonance and groupthink. In contrast, nemonik thinking aims for success, which is to obtain what you seek and to escape what you suffer. Therefore, nemonik thinking is goal oriented, which fosters freedom, alignment, compassion, allies, and win-win strategies. You might be the smartest thinker in the world, but only nemonik thinking could make you the smartest thinker you can be. This manuscript is an abridged version of *Think Smarter with Nemonik Thinking (Schade, 2016).*

Download a free eBook version
@ nemonik-thinking.org

Lao Zi's Dao De Jing

Lao Zi's Dao De Jing (Schade, 2016).

This book comprises synchronised Chinese and English versions of *Dao De Jing,* which means—*The Way of Nature and the Way of People.* Lao Zi was a Chinese philosopher who lived during the 6th century BC but is still ahead of our time. His brilliance outshines intellectual giants such as Confucius, Sun Zi, Socrates, Plato, and Aristotle. Lao Zi's aim is to teach success, which is to *obtain what you seek and escape what you suffer.* Success is achieved by aligning the *Way of People* with the *Way of Nature.* Lao Zi's success is secular and based on competence, rather than devotion. It is about positioning, rather than competing. Lao Zi's deep understanding of nature and people is crucial for your immediate survival and that of the next generation. We are facing overpopulation, dwindling resources, nuclear warfare, pollution, climate change, etc. We cannot solve those problems with the same way of thinking that is causing them. The brutal reality shows that our way of thinking is failing. Therefore, Lao Zi's eternal wisdom is the guiding light for our future. Its simplicity reaches peacefully across the boundaries of race, religion, spiritualism, ideology, and science.

Download a free eBook version

@ nemonik-thinking.org

Lao Zi Explained

Lao Zi's Dao De Jing Explained (Schade, planned 2017).

For more than two and a half thousand years, *Dao De Jing* has been shrouded in mystery. The poetic beauty of Lao Zi's words has maintained its dazzling shine that hides his esoteric secrets. In accord, Lao Zi wrote—*My words are very easy to understand and very easy to apply. Yet, people cannot understand them and they cannot apply them.* Many scholars have attempted unsuccessfully to peel away layer after layer of meaning to unravel its cryptic secrets. In contrast, the present book reveals Lao Zi's secret teachings for the first time in a clearly understandable way, imparting hidden knowledge about the *Way of Nature* and the *Way of People*. The core of Lao Zi's teachings is success, which is—*obtaining what you seek and escaping what you suffer.* Success is secular and based on competence, rather than devotion. It is about positioning, rather than competing. It is achieved by aligning the *Way of People* with the *Way of Nature*. The guiding principle for the present explanation is the core of Lao Zi's philosophy—*The One generated the Two.* Humanity is facing huge manmade problems with a failing way of thinking. Therefore, Lao Zi's eternal wisdom is more relevant than ever.

Download a free eBook version
@ nemonik-thinking.org

Lao Zi for Nemonik Thinkers

Lao Zi's Dao De Jing for Nemonik Thinkers (Schade, 2016).

Lao Zi's *Dao De Jing* has no rational sequence comprising an introduction, main body, and discussion. Even the division of the manuscript in the parts *Dao* and *De* is ambiguous. Topics concerning the *Way of Nature* and the *Way of People* appear almost ad random in the parts *Dao* and *De*. Similar to the notation *Jing*, the division in *Dao* and *De* might have been added later. Lao Zi's unfamiliar format suggests that he used a holistic, rather than a rational approach. He seems to walk around the topic, while telling the reader what he is seeing from different angles. Although that approach enhances the mystery and poetic beauty of that amazing manuscript, it did not produce the most efficient teaching tool. Therefore, I have used the nemonik template to restructure *Dao De Jing* for nemonik thinkers. This template was introduced in *Think Smarter with Nemonik Thinking (Schade, 2016).*

Lao Zi Meta-translation

Lao Zi's Dao De Jing: Meta-translation (Schade, 2016).

Lao Zi's eternal wisdom shines through in the numerous English translations of his *Dao De Jing.* Nevertheless, comparisons show that some individual Chinese pictographs and their interpretations are unclear. Therefore, this meta-translation is based on seven reputable Chinese versions. In order to select the most reliable pictographs, each one was compared across all versions. As changes might have occurred over time, the chance of a pictograph being included depended on the age of the version in which it appears. The consistent use of each pictograph was enhanced by computer assisted comparisons across the entire text. In addition, ten reputable English translations were synthesized in order to extract an initial context for each pictograph. The selected pictographs were translated with *Lao Zi's Dao De Jing: Chinese-English Dictionary (Schade, 2016)* that was compiled for this purpose. The guiding principle for this meta-translation has been the core of Lao Zi's philosophy—*The One generated the Two.* The importance of that sentence is explained in *Lao Zi's Dao De Jing Explained (Schade, 2017).*

Download a free eBook version

@ nemonik-thinking.org

Lao Zi Dictionary

Lao Zi's Dao De Jing: Chinese-English Dictionary (Schade, 2017).

Lao Zi's manuscript is more than 2,500 years old, while most Chinese-English dictionaries focus on the modern meaning of Chinese pictographs. Therefore, this special dictionary was compiled from several reputable public resources in order to get as close to the true meaning of each pictograph as possible. *Lao Zi's Dao De Jing: Meta-translation (Schade, 2016)* is based on seven Chinese versions of *Dao De Jing.* Altogether, those versions comprise about 1,600 different pictographs, which are included in the present dictionary. Furthermore, this dictionary introduces a unique numerical coding system for Chinese pictographs that could improve the search method concerning hard copies of Chinese reference books.

Download a free eBook version
@ nemonik-thinking.org

Sun Zi's The Art of War

Sun Zi's The Art of War (Schade, planned 2017).

Sun Zi (554-496 BC) was a Chinese warrior-philosopher who wrote the military classic *Bing Fa* or *The Art of War*. Although his book is about war, his strategies apply to every facet of daily life. Sun Zi deals with the art of positioning yourself in space, matter, and time. He addresses the questions raised by nemonik thinking of where, what, and when to advance, stay, retreat, accumulate, preserve, dispose, act, wait, prepare, accept, reject, reveal, and conceal. Think smarter and incorporate Sun Zi's strategies in your thinking. To become skilled in nemonik thinking, it is recommended to study—*Think Smarter with Nemonik Thinking (Schade, 2016).*

Download a free eBook version
@ nemonik-thinking.org

WEBSITE

It is the aim of my website to provide interactive on-line information about nemonik thinking. This includes discussions, books, blog, videos, exercises, updates, activities, web links, and tests. Join the nemonik thinkers and receive the latest updates. It is a work in progress. Check it out and have your say! I look forward to your feedback at:

nemonik-thinking.org

Download free eBooks and videos!

ENDNOTES

[1] The field of psychology could be divided in theoretical, experimental, and applied psychology.

[2] Data file: 161005-Vostok10KyrsGlobalDatasets.xlsm / GlobalVostok10KyrsSW|16| / cell F10033 & cell C10030 / @ http://nemonik-thinking.org/data-and-analyses.html.

[3] Dataset 9 (GlobalTempMonthly) in (Schade, Global Warming is the Solution, 2016). This dataset contains the monthly and annual global land-ocean temperature index in 0.01 degrees Celsius (°C) relative to the base period: 1951-1980, during the observation period from calendar year 1880 to 2014. Citation Dataset 9: GISTEMP Team, 2015: GISS Surface Temperature Analysis (GISTEMP). NASA Goddard Institute for Space Studies. Dataset accessed 2015-05-03 at http://data.giss.nasa.gov/gistemp/. Hansen, J., R. Ruedy, M. Sato, and K. Lo, 2010: Global surface temperature change, Rev. Geophys., 48, RG4004, doi: 10.1029/2010RG000345. Data file: Global135Yrs-TemperatureYrs.xlsm / Dataset 2.0 (GlobalRawTemp)|9|/ @ http://nemonik-thinking.org/data-and-analyses.html.

[4] Dataset 21 (Hawaiian CO2) in (Schade, Global Warming is the Solution, 2016). This dataset is based on data provided by the Mauna Loa Observatory (Scripps / NOAA / ESRL) and

comprises the Mo*nthly Mean CO2 Concentrations (ppm). The file comprises Scripps data from March 1958 to April 1974 and NOAA-ESRL data from May 1974 onwards. This dataset was released by NOAA-ESRL on September 7, 2015 and downloaded for the study on 14/09/2015 13:47:19 from http://co2now.org/Current-CO2/CO2-Now/noaa-mauna-loa-co2-data.html. Data file: HawaiiCO2(1959-2014).xlsm / HawaiiCO2 / @ http://nemonik-thinking.org/data-and-analyses.html. Due to missing data points, the observation period of the dataset was truncated from 1958-2015 to 1959-2014 (55 Years) creating data file: HawaiiCO2(1959-2014).xlsm / HawaiiCO2 (graph)/ Dataset 21 (Hawaii CO2) /. This dataset contains the following Hawaiian variables: (Year)—calendar years from 1959 to 2014; (CO2h)—Hawaiian annual atmospheric carbon dioxide concentrations in part per million (ppm).

[5] Dataset 1 (VostokDeutTempDepth423Kyrs) in (Schade, Global Warming is the Solution, 2016). This dataset contains the following Vostok variables: (Depth corrected)—ice depth from 3,310 metres to 0 metre in steps of 1 metre; (Ice Age (GT4))—ice age in years from 422,766 years BP to calendar year 1989; (deut)—depth-scale deuterium concentrations in ‰ Standard Mean Ocean Sea Water (SMOW); and (DeltaTS)—depth-scale atmospheric temperature deviations

in degrees Celsius (°C) relative to the Vostok base 1850-1989 = 0.00 °C. Citation: Petit, J.R., et al., 2001, Vostok Ice Core Data for 420,000 Years, IGBP PAGES/World Data Center for Paleoclimatology Data Contribution Series #2001-076. NOAA/NGDC Paleoclimatology Program, Boulder CO, USA. Original reference: Petit J.R., Jouzel J., Raynaud D., Barkov N.I., Barnola J.M., Basile I., Bender M., Chappellaz J., Davis J., Delaygue G., Delmotte M., Kotlya-kov V.M., Legrand M., Lipenkov V., Lorius C., Pépin L., Ritz C., Saltzman E., Stievenard M., 1999, Climate and Atmospheric History of the Past 420,000 years from the Vostok Ice Core, Antarctica, Nature, 399, pp. 429-436. Please cite the original reference when using this data. The dataset was derived from an ice core drilled by a French-Russian team at the Vostok station in Antarctica. Vostok is situated at North-bound latitude -78.47 * South-bound latitude -78.47; West-bound longitude 106.8 * East-bound longitude 106.8. The base was derived from the surface ice down to 7 metres depth. Each Ice Age (GT4) is a midpoint of a period determined by 1 metre of ice. At seven metres depth, the temporal midpoint of the corresponding metre of ice equals 129 years and at eight meters 149 years BP. In accord, the meter of ice starts at ((129 + 149) / 2) = 139 years BP. Hence, the base period ranges from 139 BP to calendar year

1989. Therefore, the base includes the period from calendar year 1850 to 1989 with a mean temperature deviation of 0.00 °C. Quotes from the NOAA website: "http://www.ncdc.noaa.gov/paleo/icecore/antarctica/vost ok/vostok_isotope.html. Downloaded Friday, 24-Jun-2011 22:33:43 EDT. Last Updated Wednesday, 20-Aug-2008 11:24:22 EDT by paleo@noaa.gov. Please see the Paleoclimatology Contact Page or the NCDC Contact Page if you have questions or comments." "NAME OF DATASET: Vostok Ice Core Data for 420,000 Years. LAST UPDATE: 11/2001 (Original Receipt by WDC Paleo). CONTRIBUTOR: Jean Robert Petit, LGGE-CNRS. IGBP PAGES/WDCA CONTRIBUTION SERIES NUMBER: 2001-076. SUGGESTED DATA CITATION: Petit, J.R., et al., 2001, Vostok Ice Core Data for 420,000 Years, IGBP PAGES/World Data Center for Paleoclimatology Data Contribution Series #2001-076. NOAA/NGDC Paleoclimatology Program, Boulder CO, USA." "PLEASE CITE ORIGINAL REFERENCE WHEN USING THIS DATA!" Data file: Vostok423Kyrs-TemperatureKyrs.xlsm / Dataset 1.0 (VostokRawTemp) |1|/ @ http://nemonik-thinking.org/data-and-analyses.html.

[6] Figure 26 (Dataset 18) in (Schade, Global Warming is the Solution, 2016). Data file: Vostok423Kyrs-nemonik-thinking.org/

TemperatureCO2.xlsm / VostokTempCO2 (Kyr423-Kyr1) |18| / @ http://nemonik-thinking.org/data-and-analyses.html. TDv is not detrended in order to facilitate the comparison with the raw CO2v values. The TDv data intervals are irregular due to glacial drift. The pressure of the ice increases with depth and. therefore, the time associated with a meter of ice increases with depth.

[7] The observation period is not exactly 420 Kyrs, but this refers to the title (Petit, Vostok Ice Core Data for 420,000 Years, 2001).

[8] Test 2 (Dataset 18) in (Schade, Global Warming is the Solution, 2016). Data file: Vostok423Kyrs-TemperatureCO2(Ranked-rpTestCorrelationTDv-CO2v) / p-value / cells B10003-10014 / @ http://nemonik-thinking.org/data-and-analyses.html.

[9] Generally, a hypothesis would be accepted if $p < 0.05$, because a 5% likelihood of making a mistake would be considered to be marginally acceptable.

[10] Data file: Vostok423Kyrs-TemperatureCO2.xlsm / Vostok-TempCO2 (Kyr423-Kyr1) / cells A4; A2-D3 / @ http://nemonik-thinking.org/data-and-analyses.html.

[11] Figure 41 (Dataset 25) in (Schade, Global Warming is the Solution, 2016). Data file: ThermalGap.xlsm / Vostok-

TempCO2 (Kyr423-Kyr11) |25|/ @ http://nemonik-thinking.org/data-and-analyses.html. The data intervals are irregular due to glacial drift. The pressure of the ice increases with depth and. therefore, the time-scale becomes non-linear.

[12] Figure 41 (Dataset 25) in (Schade, Global Warming is the Solution, 2016). Data file: ThermalGap.xlsm / Vostok-TempCO2 (Kyr423-Kyr11) |25|/ @ http://nemonik-thinking.org/data-and-analyses.html.

[13] The magnitude of the linear thermal gap at 399 ppm is already 10.1 °C, which is sufficient for the current debate. The polynomial function would further increase that thermal gap.

[14] Data file: 161005-ThermalGapData.xlsm / TDvLin / cells B2 and B4 / @ http://nemonik-thinking.org/data-and-analyses.html.

[15] Data file: 161005-ThermalGapData.xlsm / TDvLin / cells C3 and D3 / @ http://nemonik-thinking.org/data-and-analyses.html.

[16] Figure 5 (Dataset 5) in (Schade, Global Warming is the Solution., 2016). Data file: Vostok423Kyrs-TemperatureKyrs.xlsm / Fig 1.4.1 (Detrend) |5|/ @ http://nemonik-thinking.org/data-and-analyses.html..

[17] Data file: Vostok423Kyrs-TemperatureKyrs.xlsm / Thermal-Stability|5| / column B / @ http://nemonik-thinking.org/data-and-analyses.html.

[18] The thermal cooling in Antarctica equals about -2.0 °C across 400 Kyr or -0.05 °C. across 10 Kyrs. Hence, the detrending during 10 Kyrs has little or no effect on the result of the tests.

[19] Equation 4 (Dataset 5) in (Schade, Global Warming is the Solution, 2016).

[20] Table 6 (Dataset 5) in (Schade, Global Warming is the Solution, 2016). Data file: Vostok423Kyrs-TemperatureKyrs.xlsm / ThermalStability (Table)|5| / Table 6 / @ http://nemonik-thinking.org/data-and-analyses.html. For p-value see Moore, G. P., & McCabe, D. S. (2003). *Introduction to the Practice of Statistics.* New York: W. H. Freeman and Company. Table A.

[21] Table 6 (Dataset 5) in (Schade, Global Warming is the Solution, 2016). Data file: Vostok423Kyrs-TemperatureKyrs.xlsm / ThermalStability (Table)|5| / Table 6 / @ http://nemonik-thinking.org/data-and-analyses.html.

[22] n = (423 − 11) − (window = 10) + 1 = 404. Data file: Vostok423Kyrs-TemperatureKyrs.xlsm / ThermalStability|5| /

column B / @ http://nemonik-thinking.org/data-and-analyses.html.

[23] Table 3 (Dataset 5) in (Schade, Global Warming is the Solution, 2016). Data file: Vostok423Kyrs-TemperatureKyrs.xlsm / IG-duration-Tables 1.4.#|5| / Table 3 / cells B16-D30 / @ http://nemonik-thinking.org/data-and-analyses.html.

[24] Table 4 (Dataset 5) in (Schade, Global Warming is the Solution, 2016). Data file: Vostok423Kyrs-TemperatureKyrs.xlsm / IG-duration-Tables 1.4.# / Table 4 / cells B32-D40 / @ http://nemonik-thinking.org/data-and-analyses.html. For p-value see Moore, G. P., & McCabe, D. S. (2003). *Introduction to the Practice of Statistics.* New York: W. H. Freeman and Company. Table A.

[25] The calendar year 1958 comprised incomplete data and was therefore ignored.

[26] Figure 37 (Dataset 24) in (Schade, Global Warming is the Solution., 2016). Data file: CO2-Temperature1959-2014 / HawaiiCO2-GlobalTemp|24| / @ http://nemonik-thinking.org/data-and-analyses.html.

[27] Data file: CO2-Temperature1959-2014 / HawaiiCO2-GlobalTemp / @ http://nemonik-thinking.org/data-and-analyses.html.

[28] (Schade, Global Warming is the Solution, 2016)—Test 5 (Dataset 24)—randomized correlation TDg and CO2h. In which: (TDg)—Global ranked annual temperature deviation; (CO2h)—Hawaii ranked annual CO2 concentration; observation period from 1959 to 2014. At each iteration, the correlation coefficient was computed after TDg was randomized relative to CO2h. Using ranked data avoids the statistical problems associated with asymmetric frequency distributions, unequal variances, and outliers. Data file: CO2-Temperature1959-2014(Ranked-rpTestCorrelationTDg-CO2h) / p-value / cells B10011 & B10014 / @ http://nemonik-thinking.org/data-and-analyses.html.

[29] Figure 43 (Dataset 26) in (Schade, Global Warming is the Solution., 2016). Data file: ThermalGap.xlsm / HawaiiCO2-GlobalTemp|24| / @ http://nemonik-thinking.org/data-and-analyses.html.

[30] Data file: 161005-ThermalGapData.excel / TDgLin / cells B2 and B6 / @ http://nemonik-thinking.org/data-and-analyses.html.

[31] Data file: 161005-ThermalGapData.excel / TDgLin / cell C5 and D5 / @ http://nemonik-thinking.org/data-and-analyses.html.

[32] Data file: CO2-Temperature1959-2014.xlsm / HawaiianCO2-GlobalTemp|24| / cell D63 / @ http://nemonik-thinking.org/data-and-analyses.html.

[33] Vostok data interval from 324,485 to 322,582 BP. Data file: 161005-Vostok423Kyrs-TemperatureCO2.xlsm / Vostok-TempCO2(Kyr423-Kyr1)|18| / cells C247 & G247 // & //CO2-Temperature1959-2014.xlsm / HawaiianCO2-GlobalTemp|24| / cell D63 / @ http://nemonik-thinking.org/data-and-analyses.html.

[34] Regression towards the mean—if the size of a sample is increased and the mean of that larger sample moves towards the mean of the population. Therefore, extreme high means are likely to decrease, while extreme low means are likely to increase with an increase of the sample size.

[35] Data file: CO2-Temperature1959-2014.xlsm / HawaiianCO2-GlobalTemp|24| / cell D60 / @ http://nemonik-thinking.org/data-and-analyses.html.

[36] Observation period: $2014 - 1959 = 55$ years.

[37] Figure 24 (Dataset 16) in (Schade, Global Warming is the Solution, 2016). Data file: 161005-Vostok10KyrsGlobalDatasets.xlsm / GlobalVostok10KyrsSW (graph)|16| / @ http://nemonik-thinking.org/data-and-analyses.html.

[38]Data file: 161005-Vostok10KyrsGlobalDatasets.xlsm / GlobalVostok10KyrsSW |16| / cells B89 and B9979 / @ http://nemonik-thinking.org/data-and-analyses.html.

[39] All data are based on the means of 50 years sliding windows. Data file: 161005-Vostok10KyrsGlobalDatasets.xlsm / GlobalSW|15| / cells B3-B52 / & GlobalVostok10KyrsSW|16| / cell F10033 & cell C10030 / @ http://nemonik-thinking.org/data-and-analyses.html.

[40] Data file: 161005-Vostok10KyrsGlobalDatasets.xlsm / GlobalVostok10KyrsSW|16| / cell F10033 / @ http://nemonik-thinking.org/data-and-analyses.html.

[41] Data file: 161005-Vostok10KyrsGlobalDatasets.xlsm / GlobalVostok10KyrsSW|16| / cell C10033 & cell F10033 / @ http://nemonik-thinking.org/data-and-analyses.html.

[42] The statistical probability (p) indicates the percentage of Type I errors. Therefore, it is associated with the total number of decisions to be made, rather than with the total number of observations.

[43] Data file: 161005-Vostok10KyrsGlobalDatasets.xlsm / GlobalVostok10KyrsSW|16| / cell C10036 & I10036 / @ http://nemonik-thinking.org/data-and-analyses.html.

[44] Data file: 161005-Vostok10KyrsGlobalDatasets.xlsm / GlobalVostok10KyrsSW|16|/ cell H10036; I10036; F10033 / @ http://nemonik-thinking.org/data-and-analyses.html.

[45] Data file: 161005-Vostok10KyrsGlobalDatasets.xlsm / GlobalVostok10KyrsSW|16| / cell I10037 / @ http://nemonik-thinking.org/data-and-analyses.html.

[46] Data file: 161005-Vostok10KyrsGlobalDatasets.xlsm / GlobalSW(trend)|15| / cell B141 & B142 // & // 161005-Vostok10KyrsGlobalDatasets.xlsm / VostokGlobalSW(trend slope)|16| / cell B58 & B59 / @ http://nemonik-thinking.org/data-and-analyses.html.

[47] Data file: 161005-Vostok10KyrsGlobalDatasets.xlsm / VostokGlobal (table)|16| / Table 9 / cells B1-F14 / @ http://nemonik-thinking.org/data-and-analyses.html.

[48] Table 11 (Dataset 16) in (Schade, Global Warming is the Solution, 2016). Data file: 161005-Vostok10KyrsGlobalDatasets.xlsm / VostokSW(trend)|16| / cells B137 & B9978 // & //GlobalSW(trend)|15| / cells B2 & B136 // & // VostokGlobal (table) / Table 11 / cells B18-D30 / @ http://nemonik-thinking.org/data-and-analyses.html.

[49] Data file: 161005-Vostok10KyrsGlobalDatasets.xlsm / VostokGlobal (table)|16| / Table 11 / cells B18-D30 / @ http://nemonik-thinking.org/data-and-analyses.html.

[50] Table 11 (Dataset 16) in (Schade, Global Warming is the Solution, 2016). Data file: 161005-Vostok10KyrsGlobalDatasets.xlsm / VostokGlobal (table) / Table 11 / cells B18-D30 / @ http://nemonik-thinking.org/data-and-analyses.html.

[51] Figure 45 (Dataset 27) in (Schade, Global Warming is the Solution, 2016). Data file: ThermalGap.xlsm / VostokGlobalTempCO2Scatter|27| // & // TDvLin / thermal gap: cell B2 / @ http://nemonik-thinking.org/data-and-analyses.html.

[52] Figure 45 (Dataset 27) in (Schade, Global Warming is the Solution, 2016). Data file: ThermalGap.xlsm / VostokGlobalTempCO2(data)|27| / cells global B3 & B58; Vostok cells B59 & B334 / @ http://nemonik-thinking.org/data-and-analyses.html.

[53] Figure 45 (Dataset 27) in (Schade, Global Warming is the Solution, 2016). Data file: ThermalGap.xlsm / VostokGlobalTempCO2(data)|27| / cells global B3 & B58; Vostok cells B59 & B334 / @ http://nemonik-thinking.org/data-and-analyses.html.

[54] Data file: 161005-ThermalGapData.excel / TDgLin / cell B2 // & // TDvLin / cell B2 / @ http://nemonik-thinking.org/data-and-analyses.html.

[55] Data file: 161005-ThermalGapData.excel / TDgLin / cell D7 / @ http://nemonik-thinking.org/data-and-analyses.html.

[56] Data file: 161005-ThermalGapData.excel / TDgLin / cell B6 // & //TDvLin / cell A4 / @ http://nemonik-thinking.org/data-and-analyses.html.

[57] Data file: 161005-ThermalGapData.excel / TDgLin / cell D8 / @ http://nemonik-thinking.org/data-and-analyses.html.

[58] Data file: 161005-ThermalGapData.excel / TDgLin / cell D9 / @ http://nemonik-thinking.org/data-and-analyses.html.

[59] Data file: 161005-rp-test_Difference TDvLin versus TDg-Lin.xlsm / p-value / cells B10010 & B10013 / @ http://nemonik-thinking.org/data-and-analyses.html.

[60] Data file: 161005-ThermalGapData.excel / TDgLin / cells B2 and D7 / @ http://nemonik-thinking.org/data-and-analyses.html.

www.ingramcontent.com/pod-product-compliance
Lightning Source LLC
Chambersburg PA
CBHW071505200326
41519CB00019B/5873